ANIMAL GROUPS

AN ARMY OF CATERPILLARS

BY NATALIE HUMPHREY

Gareth Stevens PUBLISHING

Please visit our website, www.garethstevens.com. For a free color catalog of all our high-quality books, call toll free 1-800-542-2595 or fax 1-877-542-2596.

Library of Congress Cataloging-in-Publication Data

Names: Humphrey, Natalie, author.
Title: An army of caterpillars / Natalie Humphrey.
Description: New York : Gareth Stevens Publishing, [2022] | Series: Animal groups | Includes index. | Audience: Ages 6 | Audience: Grades K-1
Identifiers: LCCN 2020028916 (print) | LCCN 2020028917 (ebook) | ISBN 9781538263433 (library binding) | ISBN 9781538263419 (paperback) | ISBN 9781538263426 (set) | ISBN 9781538263440 (ebook)
Subjects: LCSH: Caterpillars–Juvenile literature. | Caterpillars–Behavior–Juvenile literature.
Classification: LCC QL544.2 .H855 2022 (print) | LCC QL544.2 (ebook) | DDC 595.7813/92–dc23
LC record available at https://lccn.loc.gov/2020028916
LC ebook record available at https://lccn.loc.gov/2020028917

Published in 2022 by
Gareth Stevens Publishing
111 East 14th Street, Suite 349
New York, NY 10003

Designer: Tanya Dellaccio
Editor: Kristen Nelson

Photo credits: Cover sandra standbridge/Moment/Getty Images; p. 5 Ilia Baksheev/Shutterstock.com; p. 7 Junjira Limcharoen/Shutterstock.com; p. 9 Buzzard films/Shutterstock.com; p. 11 NadyGinzburg/Shutterstock.com; p. 13 Elena Dvornikova/Shutterstock.com; p. 15 Pelevina Ksinia/Shutterstock.com; p. 17 anne-tipodees/Shutterstock.com; p. 19 Liz Weber/Shutterstock.com; p. 21 Kim Pin/Shutterstock.com.

Printed in the United States of America

CPSIA compliance information: Batch #CSGS22: For further information contact Gareth Stevens, New York, New York at 1-800-542-2595.

CONTENTS

Boldface words appear in the glossary.

Born Together

Caterpillars look like a kind of worm, but they're the larvae of butterflies or moths! Butterflies and moths lay lots of eggs and fly away. When those eggs **hatch**, the caterpillars stay together. This big group is called an army.

Caterpillar Colors

Caterpillars of the same kind will look very much the same. They may be brown or green to **camouflage** themselves in their plant homes. Others have special markings or bright colors. Some kinds of caterpillars change colors as they grow.

Army Nests

Some armies build nests. The nest is an important part of keeping the caterpillars safe from predators. The nest can even keep caterpillars safe from **parasites**! Caterpillar nests are made from special matter called silk. Caterpillars make the silk themselves.

Staying Safe and Warm

Caterpillars can't talk, but they have their own way of **communicating** with the army! When a caterpillar senses movement from a predator, they wiggle their body. This tells the army danger is nearby. Then, they can **protect** themselves.

It's hard for caterpillars to stay warm on their own. That's where the army comes in! When it's cold outside, caterpillars squish themselves together. When it's sunny, the caterpillars trap as much heat as they can. By sticking together, they can keep more heat in!

Finding Food

Some caterpillar armies build nests on leaves the caterpillars can eat. Other armies search for food! These armies will take turns looking for good spots before returning home. Some kinds of caterpillars eat small **insects** or animals.

Moving as One

When armies are moving, they follow each other in a line. As they walk, they leave a **scent** trail behind. This helps the caterpillars find their way. If the scent trail smells right, other caterpillars will sometimes join in too.

The Army Grows Up

When caterpillars are ready to become adults, they search for a good spot. Then, they **attach** themselves to a leaf or stick. They wrap themselves in their own silk, making a **chrysalis**. They become a pupa. Their body starts to change!

When the caterpillar is done changing into an adult moth or butterfly, it breaks out of its chrysalis. It flies away. It's time to start the life cycle over again. Then, another army is born!

GLOSSARY

attach: to fix to something

camouflage: colors or shapes in animals that allow them to blend with their surroundings

chrysalis: the case that protects a moth or butterfly as it's turning into an adult

communicate: to share thoughts or feelings by sound or movement

hatch: to come out of an egg

insect: a small, often winged, animal with six legs and three main body parts

parasite: a living thing that lives in, on, or with another living thing and often harms it

protect: to keep safe

scent: a smell

FOR MORE INFORMATION

BOOKS

Nilsen, Genevieve. *I See Caterpillars.* Minneapolis, MN: Jump!, Inc., 2020.

Tonkin, Rachel. *Egg to Butterfly.* New York, NY: Crabtree Publishing Company, 2020.

WEBSITES

The Butterfly Life Cycle

www.natgeokids.com/uk/discover/animals/insects/butterfly-life-cycle/

See pictures and read all about how an egg becomes a caterpillar, and a caterpillar becomes a butterfly.

Identify Caterpillars

www.wildlifetrusts.org/wildlife/how-identify/identify-caterpillars

Do you have caterpillars in your yard? Find out what kind here!

INDEX